RISING STARS
Mathematics

Year 1

Practice Book C

Author: Paul Broadbent

ISBN: 978-1-78339-812-6
Text, design and layout © Rising Stars UK Ltd 2015

First published in 2015 by
Rising Stars UK Ltd, part of Hodder Education Group,
An Hachette UK Company
Carmelite House
50 Victoria Embankment
London EC4Y 0DZ
www.risingstars-uk.com
Reprinted in 2016

Author: Paul Broadbent
Programme consultants: Cherri Moseley, Caroline Clissold, Paul Broadbent
Publishers: Fiona Lazenby and Alexandra Riley
Editorial: Aidan Gill, Denise Moulton
Answer checker: Deborah Dobson
Project manager: Sue Walton

Series and character design: Steve Evans
Text design: Words & Pictures
Illustrations by Steve Evans

Cover design: Steve Evans and Words & Pictures

Printed by Hobbs the Printers Ltd, Totton, Hampshire SO40 3WX
A catalogue record for this title is available from the British Library.

Contents

10a 3-D shapes and towers

1 Write the names of these shapes. Choose from this list.

cuboid	cylinder
cube	sphere
pyramid	cone

a

d

b

e

c

f

2 Sort these shapes. Write each letter in the correct box.

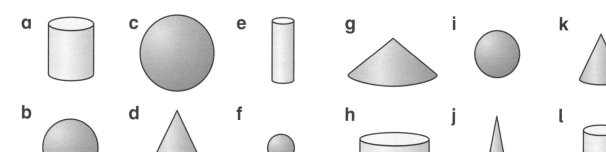

Cylinder	Sphere	Cone

 3 One shape is the odd one out in each set. Write its **name**.

Choose from these shape names:

| cone | cuboid | pyramid | cylinder | sphere |

a [blank box]

b [blank box]

c [blank box]

d [blank box]

e [blank box]

Colour the shapes that have **any curved faces**.

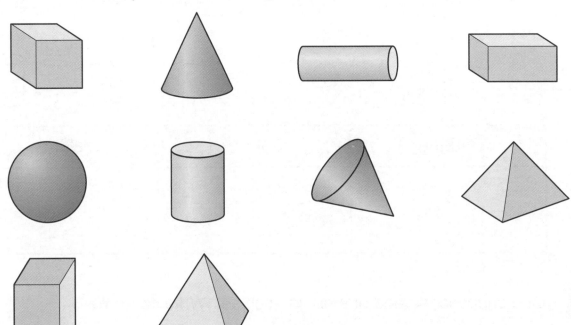

5 Match each lid to the correct box.

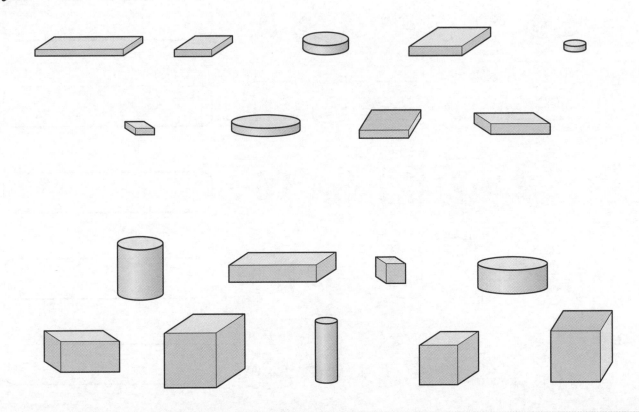

YOU WILL NEED:
- **various 3-D shape models and solid objects**

Use a mix of any of the 3-D shapes to build towers.

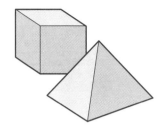

- Make each tower as tall as you can.
- Measure the height in centimetres.
- Record your results on this table.

3-D shapes used	Picture of tower	Height of tower (cm)

Which shapes made the tallest tower?

YOU WILL NEED:
• ruler

Draw **3** different shapes in each column. Use a ruler.

3 sides	4 sides

My 3-sided shape at the top is a

My 4-sided shape in the middle is a

My 4-sided shape at the bottom is a

2 Sort these 3-D shapes. Write the letters in the correct part of the Venn diagram.

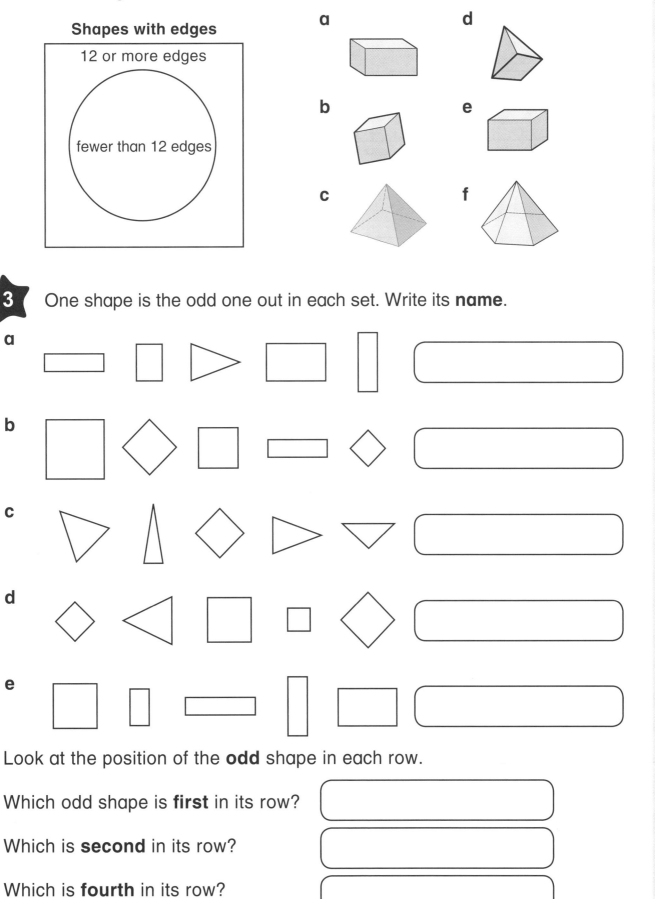

Shapes with edges

12 or more edges

fewer than 12 edges

a   d

b   e

c   f

3 One shape is the odd one out in each set. Write its **name**.

a

b

c

d

e

Look at the position of the **odd** shape in each row.

Which odd shape is **first** in its row?

Which is **second** in its row?

Which is **fourth** in its row?

4 Draw **3** more shapes to continue these patterns.

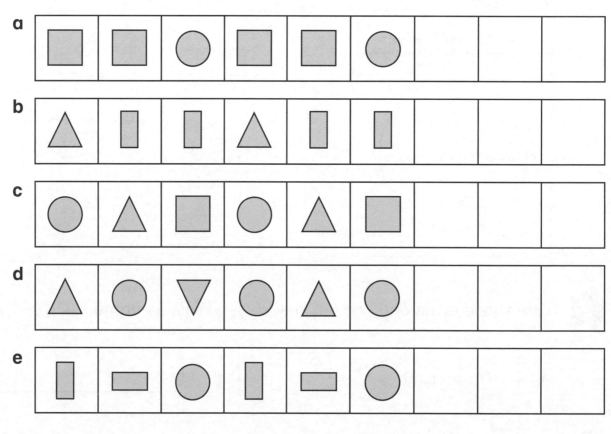

a

b

c

d

e

5 Draw your own repeating shape pattern here.

6

a Colour the shapes:

- **blue**   the triangle below a rectangle
- **pink**   the square below a circle
- **purple**   the triangle to the left of a circle
- **yellow**   the square to the right of a circle
- **red**   the circle 2 places above a triangle
- **orange**   the rectangle 2 places left of a circle
- **green**   the rectangle 2 places below a triangle
- **brown**   the rectangle 2 places right of a circle

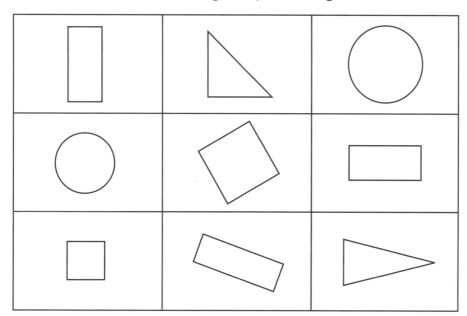

b Complete these instructions. One shape is not coloured.
Where is it?

From the large red circle it is [1] down, [2] left.

From the purple triangle it is [ ] down, [ ] left.

From the blue triangle it is [ ] left, [ ] up.

> Choose a different shape.
> How many different ways can you describe its position?

11 ★

11a Ordering

1 Write the missing numbers.

a

	40	50		70	80	

b

10	20		40		60	

c

	70	60			30	20

d

90		70		50	40	

2 Write the value of these numbers.

27 →

a 34 →

b 53 →

c

59 →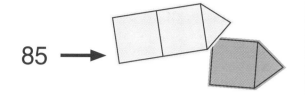

e

85 →

d

68 →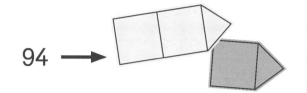

f

94 →

3 Complete these.

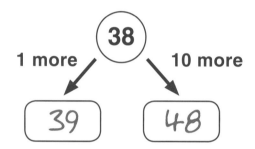

38

1 more → 39

10 more → 48

a

27

1 more

10 more

d

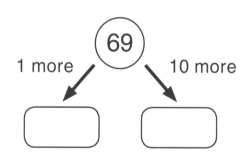

69

1 more

10 more

b

45

1 more

10 more

e

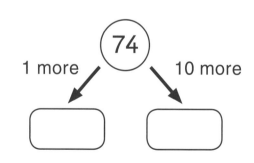

74

1 more

10 more

c

63

1 more

10 more

f

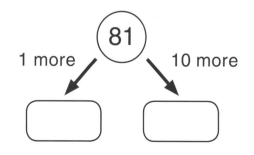

81

1 more

10 more

4 Write each set of numbers in order. Start with the **smallest** number.

a

smallest

62   19   84   56   25

b

smallest

93   44   97   75   70

c

smallest

28   43   34   23   82

d

smallest

60   68   61   16   86

 5 These 6 snails are having a race. Decide which snail is first, second, third, fourth, fifth and sixth. Write the position that each finishes.

Snail	Position in race

1 Count in fives. Find the total number of fingers in each picture.

a 3 hands waving

[____] fingers

c 5 hands waving

[____] fingers

b 4 hands waving

[____] fingers

d 6 hands waving

[____] fingers

 2 **YOU WILL NEED:**
• **different coloured crayons**

Start at 10.
• Count on in tens.
• Circle the numbers you land on.

Now start at 5.
• Count on in fives.
• Colour the numbers you land on.

What do you notice?

1	2	3	4	5	6	7	8	9	10
11	12	13	14	15	16	17	18	19	20
21	22	23	24	25	26	27	28	29	30
31	32	33	34	35	36	37	38	39	40
41	42	43	44	45	46	47	48	49	50
51	52	53	54	55	56	57	58	59	60
61	62	63	64	65	66	67	68	69	70
71	72	73	74	75	76	77	78	79	80
81	82	83	84	85	86	87	88	89	90
91	92	93	94	95	96	97	98	99	100

3 Answer these.

a   5 **more** than 35 is [ ]

e   5 **less** than 50 is [ ]

b   5 **more** than 60 is [ ]

f   5 **less** than 95 is [ ]

c   5 **more** than 85 is [ ]

g   5 **less** than 15 is [ ]

d   5 **more** than 40 is [ ]

h   5 **less** than 70 is [ ]

4 Write the missing numbers.

a

	35	40	45		

b

5		15	20		30	

c

	70	65		55		45

d

95		85		75	70	

5 Draw the 5p coins you need to pay for each.
Write the number of coins.

a

35p

Number of coins

b

40p

Number of coins

c

Colouring Book

55p

Number of coins

d

65p

Number of coins

e

70p

Number of coins

f

80p

Number of coins

Tick (✓) the shapes that show half a circle.

a ☐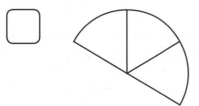

e ☐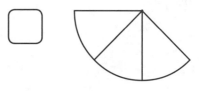

b ☐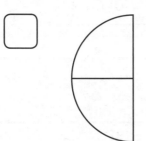

f ☐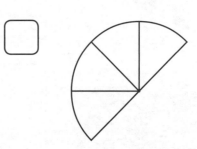

c ☐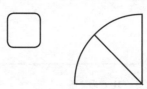

g ☐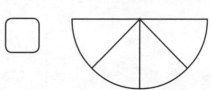

d ☐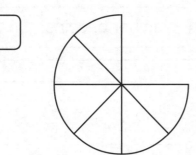

h ☐

Write the times.

a

half past [ ]

e

half past [ ]

i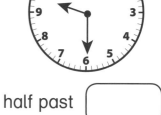

half past [ ]

b

half past [ ]

f

half past [ ]

j

half past [ ]

c

half past [ ]

g

half past [ ]

k

half past [ ]

d

half past [ ]

h

half past [ ]

l

half past [ ]

Draw hands on the clocks to show the times.

a

half past four

c

half past two

e

half past twelve

b

half past seven

d

half past eleven

f

half past three

 Join these clocks in order. Start at 2 o'clock. Finish at half past 7.

5 Draw the hands on each clock to show your answers.

a What time is half an hour later than 4 o'clock?

d What time is 1 hour earlier than half past 4?

b What time is 1 hour later than half past 6?

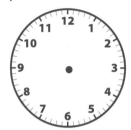

e What time is half an hour earlier than 9 o'clock?

c What time is half an hour later than 10 o'clock?

f What time is 1 hour earlier than half past 1?

 6 Join the clocks that show the same times.

1 Draw the next **3** shapes for each necklace.

a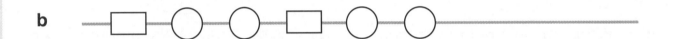

b

c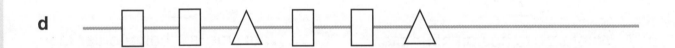

d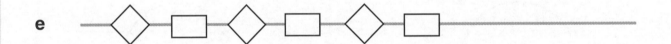

e

f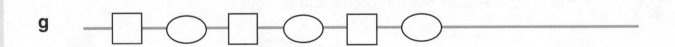

g

h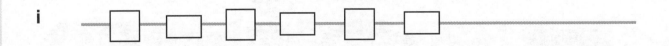

i

YOU WILL NEED:
- **2 different coloured crayons**

Colour the beads to make a repeating pattern. Use **2** different colours.
Make each necklace different.

a

b

c

3 Tick (✓) the shape that is missing from each pattern.

a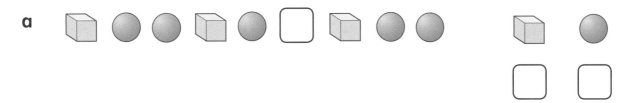

b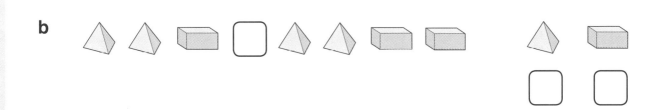

c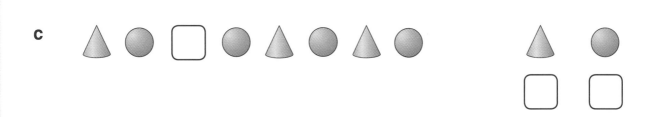

d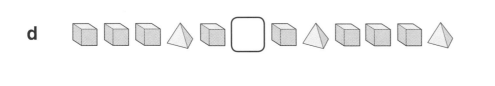

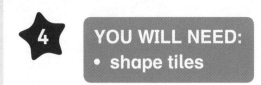

YOU WILL NEED:
• shape tiles

Use small shape tiles to make some repeating patterns on these grids. Use 2 or 3 different shapes.

When you have made a good pattern, draw it on the grid.

Continue the patterns.

a

| X | X | O | X | X | O | X | | | |

b

| 1 | 2 | 3 | 1 | 2 | 3 | 1 | 2 | 3 | 1 | | | |

c

| A | B | B | A | B | B | A | B | | | |

d

| ↓ | ↓ | ↑ | ↑ | ↓ | ↓ | ↑ | ↑ | | | |

e

| D | C | D | C | D | C | D | C | D | | | |

f

| X | 6 | 6 | X | 9 | 9 | X | 6 | 6 | X | | | |

Solving problems

12a Solving addition problems

1 Complete these. Two have been done for you.

5	**6**	**7**	**8**
0 + [5]	0 + [ ]	0 + [ ]	0 + [ ]
1 + [4]	1 + [ ]	1 + [ ]	1 + [ ]
2 + [ ]	2 + [ ]	2 + [ ]	2 + [ ]
3 + [ ]	3 + [ ]	3 + [ ]	3 + [ ]
4 + [ ]	4 + [ ]	4 + [ ]	4 + [ ]
5 + [ ]	5 + [ ]	5 + [ ]	5 + [ ]
	6 + [ ]	6 + [ ]	6 + [ ]
		7 + [ ]	7 + [ ]
			8 + [ ]

YOU WILL NEED:
• **interlocking cubes**

Use cubes to help answer these. Write the numbers in the bar model.

6 + 3 = (9)

9	
6	3

a 4 + 7 = ( )

9 + 2 = ( )

b 7 + 5 = ( )

6 + 6 = ( )

c 8 + 5 = ( )

6 + 7 = ( )

d 9 + 5 = ( )

7 + 7 = ( )

3 Answer these problems. Use the bar model to help you.

a There are 6 sparrows and 5 robins on a bird table. How many birds are on the table altogether?

b A plate has 8 chocolate biscuits and 4 orange biscuits. How many biscuits are on the plate?

c There are 9 red T-shirts and 7 blue T-shirts on a washing line. How many T-shirts are on the washing line?

d There are 7 maths books and 8 science books on a shelf. How many books are on the shelf?

Use the bar model and some coins to help you.

a There are 15 coins in a jar. Some are 1p coins and some are 2p coins.
How much could there be in the jar?
Complete the chart.

15 coins	

1p coins	2p coins	Total amount
15	0	15p
14	1	16p
13		
1	14	29p
0	15	30p

b There is **1 more** 2p coin than 1p coins in the jar.

How much is there altogether in the jar? [   ] p

1 Find the difference between these numbers.
Use the number line to help.

0 1 2 3 4 5 6 7 8 9 10 11 12 13 14 15

a

4 ⟶ 8

b

4 ⟶ 9

c

5 ⟶ 9

d

5 ⟶ 10

e

6 ⟶ 10

f

6 ⟶ 11

g

6 ⟶ 12

h

7 ⟶ 12

2 Write the 4 maths facts for each of these.

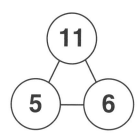

$5 + \boxed{6} = 11$ $11 - \boxed{5} = 6$

$\boxed{6} + 5 = \boxed{11}$ $\boxed{11} - \boxed{6} = 5$

a

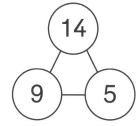

$9 + \boxed{} = 14$ $14 - \boxed{} = 9$

$\boxed{} + 9 = \boxed{}$ $\boxed{} - \boxed{} = 5$

b

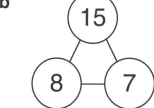

$\boxed{} + 7 = \boxed{}$ $\boxed{} - \boxed{} = 8$

$\boxed{} + 8 = \boxed{}$ $\boxed{} - \boxed{} = 7$

c

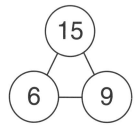

$\boxed{} + \boxed{} = \boxed{}$ $\boxed{} - \boxed{} = \boxed{}$

$\boxed{} + \boxed{} = \boxed{}$ $\boxed{} - \boxed{} = \boxed{}$

d

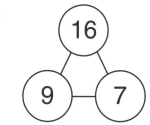

$\boxed{} + \boxed{} = \boxed{}$ $\boxed{} - \boxed{} = \boxed{}$

$\boxed{} + \boxed{} = \boxed{}$ $\boxed{} - \boxed{} = \boxed{}$

YOU WILL NEED:
- **interlocking cubes**

Use cubes to help answer these. Write the numbers in the bar model.

9	
3	6

9 – 3 = (6)

a

8	
6	

8 – 6 = ( )

e

12	
9	

12 – 9 = ( )

i

16 – 4 = ( )

b

9	
4	

9 – 4 = ( )

f

14	
7	

14 – 7 = ( )

j

19 – 5 = ( )

c

10	
3	

10 – 3 = ( )

g

15	
6	

15 – 6 = ( )

k

13 – 2 = ( )

d

11	
5	

11 – 5 = ( )

h

18	
8	

18 – 8 = ( )

l

17 – 10 = ( )

4 Answer these problems. Use the bar model to help you.

a 15 balloons were blown up and ready for a party. After the party only 6 balloons had **not** burst. How many balloons burst during the party?

b There are 16 sandwiches on a plate. 10 are cheese and the rest are egg. How many egg sandwiches are on the plate?

c A cake has 10 candles. So far I have lit 5 candles. How **many more** candles do I need to light?

d We ate 8 sausages from a string of 12 sausages. How many sausages were left?

e Invitations were sent to 14 children. Five were unable to come to the party. How many children came to the party?

5

Answer these. Lay out counters and cubes in the grid to help.

There are **3 more** counters than cubes.

If there are 5 counters, how many cubes are there? 2

a There are **5 more** counters than cubes.
If there are 9 counters, how many cubes are there? cubes

b There are **2 more** cubes than counters.
If there are 8 cubes, how many counters are there? counters

c There are **3 fewer** counters than cubes.
If there are 4 counters, how many cubes are there? cubes

d There are **4 fewer** cubes than counters.
If there are 6 cubes, how many counters are there? counters

Try this tricky problem.

e There are **3 more** counters than cubes. Altogether there are
13 counters and cubes. How many are there of each?

There are cubes and counters

YOU WILL NEED:
* cubes
* number cards 1–10

Solve this problem. Use cubes to help you.

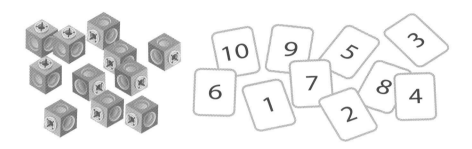

a I have some number cards to 10. I choose 2 of them. When I add the 2 numbers together they total 11. What could my 2 numbers be?

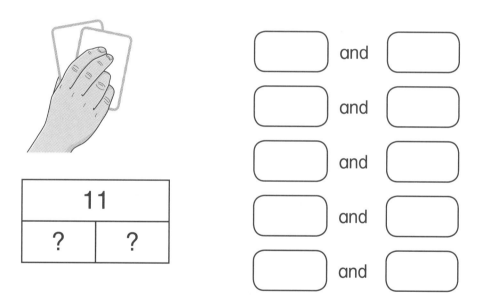

11	
?	?

◯ and ◯

◯ and ◯

◯ and ◯

◯ and ◯

◯ and ◯

b The difference between the 2 numbers on my cards is 5. What are my 2 numbers?

?	
?	5

◯ and ◯

Exploring halves, quarters and arrays

13a Halves

 1 Draw **double** the number of cherries.

Write how many cherries are on each cake.

a

d

b

e

c

f

2 What **total** do these domino doubles show?

a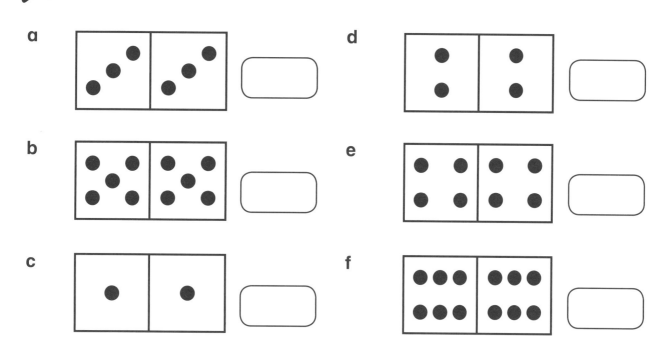

b

c

d

e

f

3 Circle the **whole** objects. Join the matching halves.

Tick (✓) the shapes that show **equal halves**.

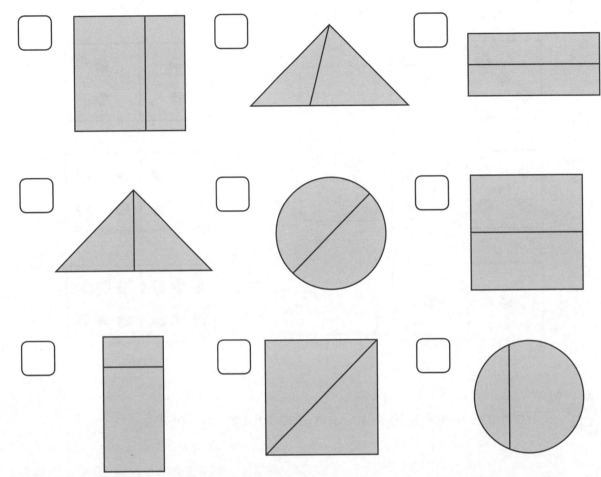

YOU WILL NEED:
- **coloured crayons**
- **ruler**

Draw a line to **halve** each shape. Colour half of each shape.

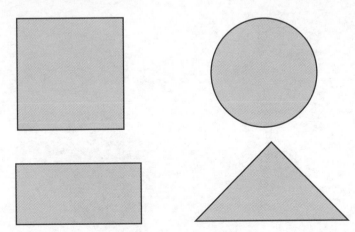

Draw a loop around **half** the stars. Complete each sentence.

a

Half of 6 is [ ]

c

Half of 10 is [ ]

b

Half of 8 is [ ]

d

Half of 12 is [ ]

 YOU WILL NEED:
• **coloured crayons or pencils**

Colour **half** of each flag. Make each flag different.

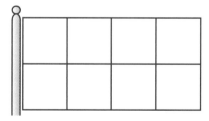

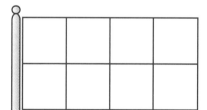

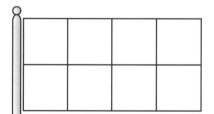

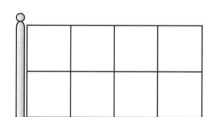

1 Tick (✓) the shapes that show **equal quarters**.

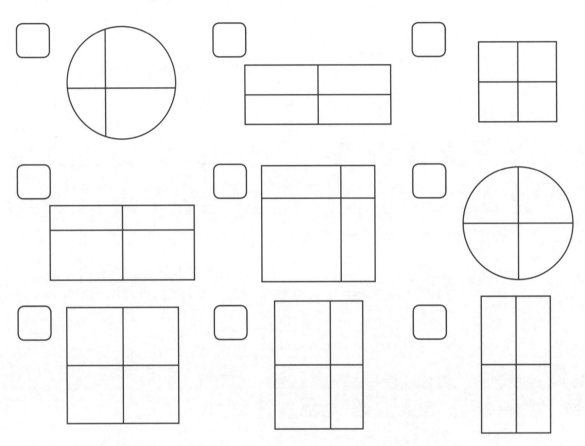

2 Draw lines to **quarter** each pizza.

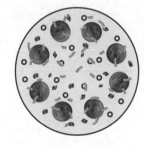 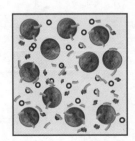 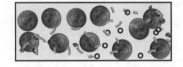 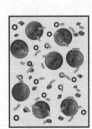

3 Each tray has been divided into quarters.
Count **1 group** to find how many are in **one quarter**.

a

One quarter of [4] is [ ]

b

One quarter of [8] is [ ]

c

One quarter of [ ] is [ ]

d

One quarter of [ ] is [ ]

e

One quarter of [ ] is [ ]

4

YOU WILL NEED:
• **interlocking cubes**

Use cubes to solve this problem.

You have 12 cubes. They are red, blue and white.
Half the cubes are red. One quarter of the cubes are blue.

How many of the cubes are white? [ ]

1 Write the next 4 numbers in each pattern.

a 2 4 6 8 10 12 ☐ ☐ ☐ ☐

b 5 10 15 20 25 30 ☐ ☐ ☐ ☐

c 10 20 30 40 50 60 ☐ ☐ ☐ ☐

 2 Count these. Write the answers.

a

☐ + ☐ + ☐ + ☐ + ☐ = ☐

5 groups of 3 = ☐

b

☐ + ☐ + ☐ + ☐ = ☐

4 groups of 2 = ☐

c

☐ + ☐ + ☐ = ☐

3 groups of 4 = ☐

d

☐ + ☐ + ☐ + ☐ + ☐ = ☐

5 groups of 4 = ☐

3 How many eggs are in each box? Count them in groups.

a

☐ eggs

c

☐ eggs

e

☐ eggs

b

☐ eggs

d

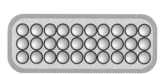

☐ eggs

f

☐ eggs

4 Freda the frog has hopped along the lily pad in twos.

How many jumps of **2** has Freda made to reach **8**?

$$\boxed{4}$$ jumps of **2** equals **8**.

Look at these different jumps. Complete the number sentences.

a

How many jumps of **5** has Freda made to reach **10**?

$$\boxed{}$$ jumps of **5** equals **10**.

b

How many jumps of **3** has Freda made to reach **12**?

$$\boxed{}$$ jumps of **3** equals **12**.

c

How many jumps of **2** has Freda made to reach **12**?

$$\boxed{}$$ jumps of **2** equals **12**.

a Draw loops to show
4 groups of 5.

b Draw loops to show
5 groups of 4.

What is the same and what is different about your 2 arrays?

14a Different turns

YOU WILL NEED:
- coloured crayons

Find the shapes that are divided into **halves**. Colour one half **red**.

Find the shapes that are divided into **quarters**. Colour one quarter **blue**.

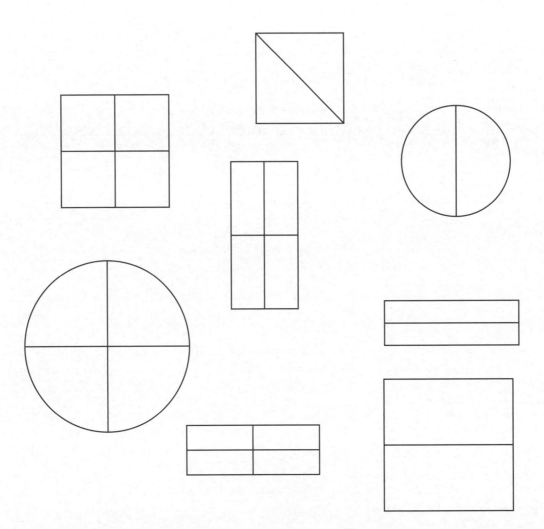

Which parts of the shapes show **three-quarters**? Colour these **yellow**.

2 Look at these cars on roads.

Draw arrows to show which direction they will go – **left** or **right**.

a

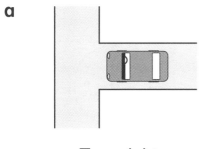

Turn right

c

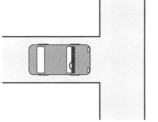

Turn left

e

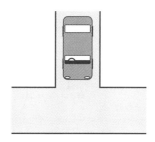

Turn right

b

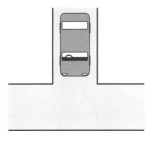

Turn left

d

Turn right

f

Turn left

3 Look at the boats and their turns.

a

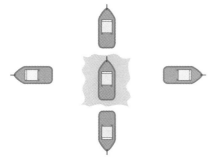

Circle the boat that has made a **half** turn to the **right**.

c

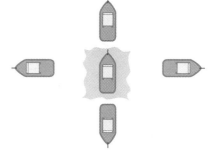

Circle the boat that has made a **quarter** turn to the **left.**

b

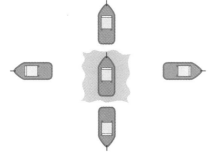

Circle the boat that has made a **quarter** turn to the **right**.

d

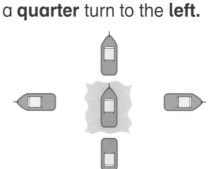

Circle the boat that has made a **whole** turn to the **right**.

Draw an arrow to show the turns.

a

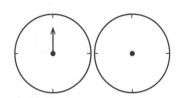

Show a **quarter** turn **right**.

c

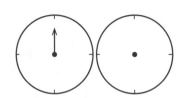

Show a **three-quarter** turn **left**.

e

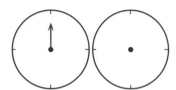

Show a **quarter** turn **left**.

b

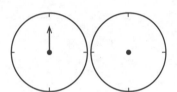

Show a **half** turn.

d

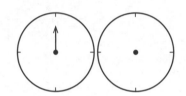

Show a **full** turn.

f

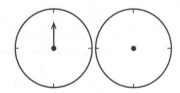

Show a **three-quarter** turn **right**.

YOU WILL NEED:
- **different shape tiles**

Choose a triangle shape tile.

Place the triangle in the first box below. Draw round it.

Now rotate the triangle a quarter turn in the next box. Do the same for each box.

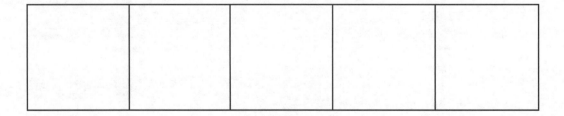

Draw your own grid on paper. Then try this with different shapes.

 1 The robot is facing the house. It follows the instructions to turn.
What will the robot turn to face each time?
After each turn the robot returns to its start position, facing the house.

house

bear

car

boat

a

Half turn to
the **right.**

c

Three-quarter
turn to the **right.**

e

Full turn to
the **left.**

b

Quarter turn to
the **left.**

d

Quarter turn
to **right.**

f

Three-quarter
turn to **the left.**

2 YOU WILL NEED:
• counters

Place a counter on the **START** box.

Follow the routes for Amy and Piara on this map.

What place will each girl visit?

Follow the shaded roads only.

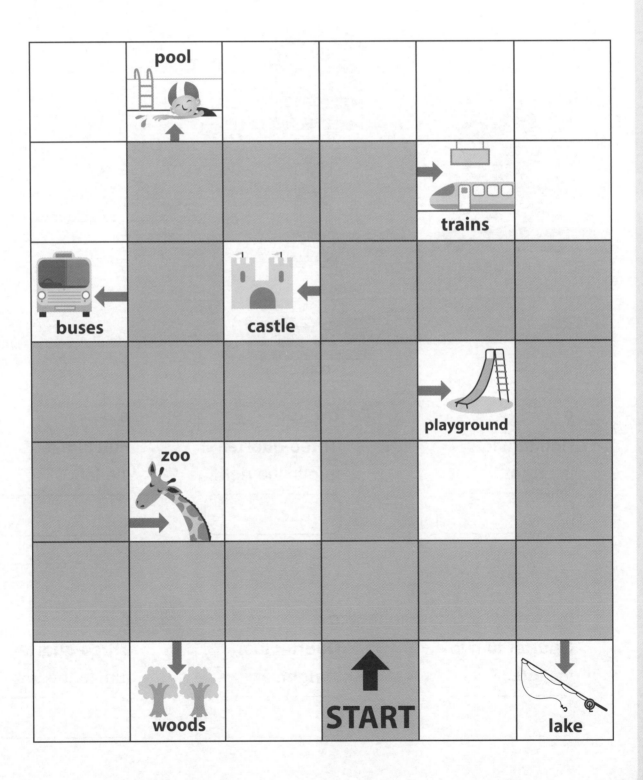

a   Amy's day out

From the START box, go:

Forward 1
Quarter turn left

Forward 3
Quarter turn right

Forward 2
Quarter turn right

Forward 1
Quarter turn left

Forward 2
Quarter turn right

Forward 1
Quarter turn right

Where is Amy visiting?

b   Piara's day out

From the START box, go:

Forward 1
Quarter turn right

Forward 2
Quarter turn left

Forward 3
Quarter turn left

Forward 2
Quarter turn left

Forward 1
Quarter turn right

Forward 3
Quarter turn left

Forward 1
Quarter turn left

Where is Piara visiting?

c   Can you find a shorter route for Piara's journey?

d   Choose a place to visit.

Write a route to get there from the **START.**

e   A bus takes passengers to **all** the places to visit.

Can you plan a bus route with a bus stop at each place?

3 Describe this route from START to FINISH.

Use these instructions:

forwards   backwards   quarter turn   left   right

RISING STARS
Mathematics

Rising Stars Mathematics includes high-quality Textbooks, Teacher's Guides, Practice Books, online tools and CPD to provide a comprehensive mastery programme for mathematics.

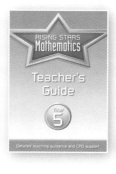

For more information on the complete range, visit
www.risingstars-uk.com/rsmathematics.

RISING STARS
Mathematics

Practise all
the skills you
have learnt
in class with
Jen and Sam!

Year
1 C

ISBN 978-1-78339-812-6

9 781783 398126 >

RISING ★ STARS

For more information please call 0800 091 1602
www.risingstars-uk.com

Follows
the NCETM
textbook
guidance